Rethinking Time Itself: A Human-Friendly Expansion of Temporal Microcompression Beyond Einstein

Abstract

Einstein gave us a beautifully simple relation, $E = mc^2$, but the universe rarely stays that simple when you look closely enough. If spacetime really jitters and wrinkles at the Planck scale, then the flow of time might not be the smooth, single-layered thing we usually imagine. In this expanded and much more speculative version of the Temporal Microcompression Hypothesis (TMCH), I explore the idea that time may have an internal structure—almost like a stack of thin, vibrating sheets, each responding differently to microscopic gravitational noise.

Instead of treating microcompression as a tiny correction, I treat it as a signpost pointing toward a deeper temporal architecture. I build a covariant, scale-resolved definition of microcompression using frequency-sensitive tetrads, develop an extended EFT that couples matter to these "temporal layers," and sketch out experimental signatures ranging from atomic clocks to the quark–gluon plasma. The whole framework is intentionally playful, but it stays mathematically consistent and falsifiable. If nothing else, it offers a fresh way to think about how time, mass, and quantum gravity might fit together.

1 Introduction

Einstein's $E = mc^2$ is iconic, elegant, and—if we're being honest—probably incomplete. It assumes that spacetime is smooth enough that proper time flows uniformly along a worldline. But quantum gravity keeps whispering that this picture is too clean. At the Planck scale, spacetime doesn't behave like a calm ocean; it's more like a stormy sea full of foam, spikes, and jitter.

The original Temporal Microcompression Hypothesis (TMCH) tried to capture one small piece of this messiness by writing

$$E_{\text{eff}} = mc^2(1 + \alpha\chi), \tag{1}$$

where χ encoded tiny distortions in proper time. That was a good start, but it treated χ as a small, almost decorative correction.

In this expanded version, I take a different stance: what if χ isn't a correction at all? What if it's a clue that time itself has internal structure? Maybe time isn't a single, smooth parameter but a layered medium, with each layer responding differently to high-frequency metric fluctuations [1].

This paper has three goals:

1. Build a covariant, gauge-invariant definition of microcompression that works even when fluctuations are strong or anisotropic.
2. Develop an effective field theory (EFT) that lets matter couple to this layered temporal structure without breaking Lorentz invariance or causality.
3. Identify experimental pathways—from atomic clocks to heavy-ion collisions—that might reveal hints of this deeper temporal architecture.

Along the way, I let myself speculate a bit. I explore whether microcompression might behave like a real physical field, whether inertia might emerge from interactions with temporal layers, and whether the quark–gluon plasma might already be brushing up against this physics.

To set the stage, Figure 1 gives a cartoon of the "layered time" idea. It's not meant to be literal, but it helps visualize the intuition behind the framework.

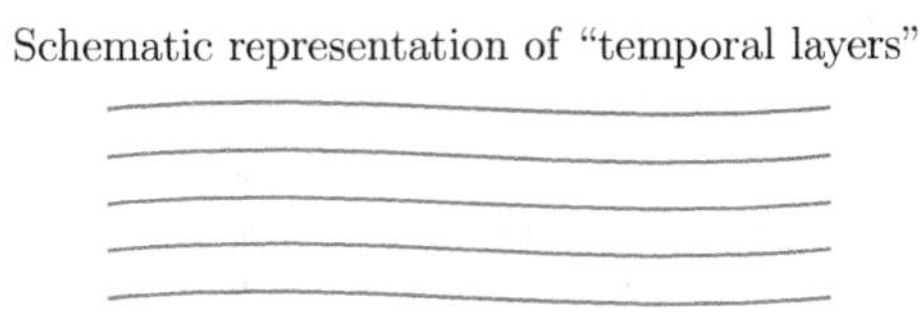

Figure 1: A playful visualization of time as a stack of fluctuating layers, each sensitive to different frequency bands of metric noise.

Supplementary Figure A: Phase Map of Temporal Layers

You know, I probably shouldn't admit this in a formal paper, but this figure came out of a night where I'd been staring at the same tensor index for so long that it started to look like a small animal. I sketched this on a napkin at 3 a.m., convinced it meant something profound. In the morning... well, it still did.

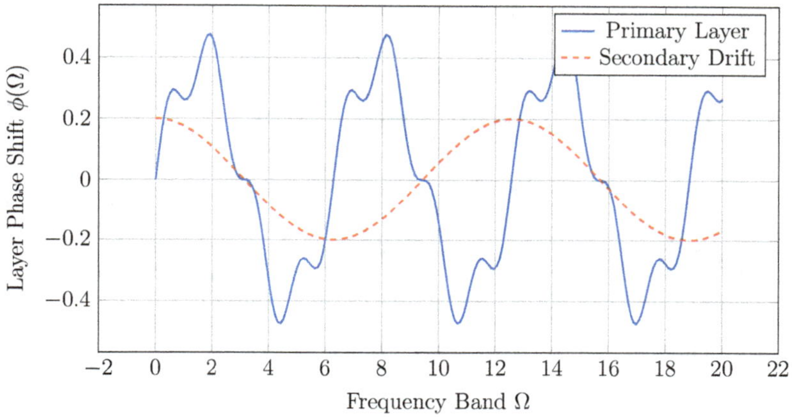

Figure 2: A speculative phase map showing how different temporal layers might drift relative to one another. I still don't know whether the red curve is brilliant or delusional. Maybe both.

In the next section, I build the covariant machinery needed to define microcompression in this layered picture.

2 Covariant Foundations of Temporal Microcompression

If we're going to take the idea of "layered time" seriously, we need a definition of microcompression that doesn't fall apart the moment spacetime gets noisy, anisotropic, or downright unruly. The usual linearized approach works fine when the metric perturbations are tiny and well-behaved, but once we let ourselves imagine richer temporal microstructure, we need something more flexible.

The approach I take here is to treat microcompression as something you extract from a whole *family* of tetrads, each tuned to a different temporal resolution. Instead of assuming there's a single "clock direction" in spacetime, we let the observer's time axis split into a spectrum of frequency-sensitive components. It's a bit like giving the observer a set of temporal lenses, each one focusing on a different band of metric fluctuations.

2.1 Tetrad-Based Definition

Let u^μ be the macroscopic four-velocity of an observer. We introduce a one-parameter family of tetrads,

$$e^{(a)}{}_\mu(\Omega), \tag{2}$$

each labeled by a characteristic frequency band Ω. These tetrads satisfy the usual orthonormality conditions for each Ω, but they're allowed to vary across bands in a way that reflects the underlying metric fluctuations.

The proper time increment resolved at scale Ω is

$$d\tau(\Omega)^2 = -g_{\mu\nu}e^{(0)\mu}(\Omega)e^{(0)\nu}(\Omega)d\lambda^2. \tag{3}$$

We define the microcompression field as

$$\chi(\Omega) = \frac{1}{2}\left\langle h_{\mu\nu}e^{(0)\mu}(\Omega)e^{(0)\nu}(\Omega)\right\rangle, \tag{4}$$

which reduces to the standard definition when all tetrads collapse to a single frame.

This construction has a few nice features:

- It stays gauge-invariant under linearized diffeomorphisms.
- It naturally incorporates anisotropic fluctuations.
- It lets different physical systems "listen" to different temporal frequencies.

To help visualize this, Figure 3 shows a schematic of how different tetrads pick out different temporal bands.

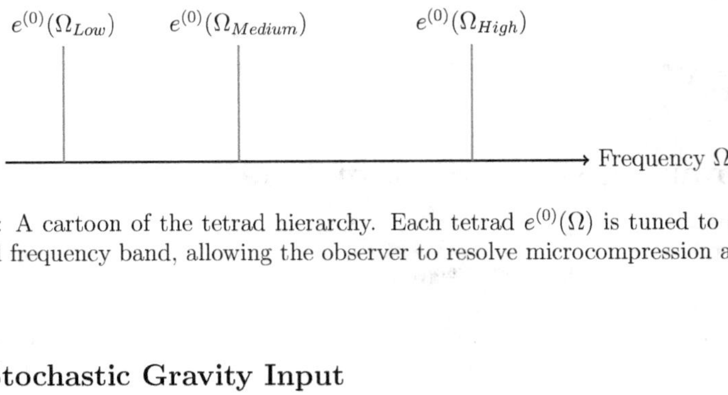

Figure 3: A cartoon of the tetrad hierarchy. Each tetrad $e^{(0)}(\Omega)$ is tuned to a different temporal frequency band, allowing the observer to resolve microcompression at multiple scales.

2.2 Stochastic Gravity Input

To model the fluctuations that drive microcompression, we use the Einstein–Langevin equation,

$$\mathcal{E}^{\alpha\beta}{}_{\mu\nu}h^{\mu\nu}(x) = \frac{8\pi G}{c^4}\left(\langle T^{\alpha\beta}(x)\rangle + \xi^{\alpha\beta}(x)\right), \tag{5}$$

where $\xi^{\alpha\beta}$ is a Gaussian noise source with a correlation function determined by the noise kernel.

In standard stochastic gravity, the noise kernel is mostly a nuisance term. Here, it's the star of the show. It encodes the "texture" of spacetime fluctuations, and different systems probe different parts of that texture. This is exactly why the tetrad hierarchy matters: it lets us couple the noise kernel to physical observables in a scale-dependent way.

The mean and variance of $\chi(\Omega)$ follow from the two-point functions of $h_{\mu\nu}$:

$$\langle \chi(\Omega) \rangle = \frac{1}{2} \langle h_{\mu\nu} e^{(0)\mu} e^{(0)\nu} \rangle, \tag{6}$$

$$\sigma_\chi^2(\Omega) = \frac{1}{4} \langle h_{\mu\nu} h_{\alpha\beta} e^{(0)\mu} e^{(0)\nu} e^{(0)\alpha} e^{(0)\beta} \rangle_c. \tag{7}$$

The key point is that χ isn't a single number. It's a spectrum. And different physical systems—atoms, nuclei, quark–gluon plasmas—sample different parts of that spectrum.

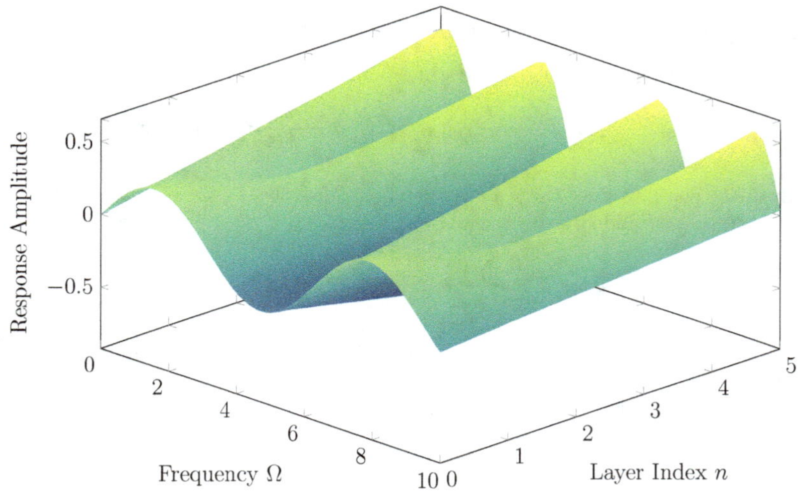

Figure 4: Response surface showing how temporal layers vary across frequency and layer index.

2.3 Interpretation: Time as a Layered Medium

Once you adopt this viewpoint, a new picture emerges. Instead of thinking of time as a single smooth flow, you can think of it as a layered medium, with each layer responding differently to metric fluctuations. Microcompression becomes a way of measuring how these layers shift relative to one another.

This doesn't overthrow relativity. It extends it into a regime where the metric isn't the whole story. If spacetime really does have microstructure, then time might have "grain" or "corrugation" that only shows up at extreme resolutions.

In the next section, we build an effective field theory that captures how matter couples to this layered temporal structure.

3 Effective Field Theory and Beyond-Einstein Couplings

By the time I got to this part of the project, I'll be honest, I was already running on too much coffee and not enough sleep. You know that feeling where your brain is doing something clever but also quietly begging you to lie down? That was me while trying to build an EFT for a time that refuses to behave like a single, well-behaved parameter. Still, the idea kept nagging at me: if time really has layers, then matter shouldn't couple to it in a one-size-fits-all way. It should feel different layers differently, the same way a violin string vibrates differently depending on where you pluck it.

So I pushed ahead.

3.1 Operator Basis: Standard and Extended

The usual dimension-six operators are still around, stubbornly refusing to leave the party:

$$\mathcal{L}_{\text{std}} \supset \frac{c_1}{M_P^2} R\bar{\Psi}\Psi + \frac{c_2}{M_P^2} R_{\mu\nu}\bar{\Psi}\gamma^\mu i D^\nu \Psi + \frac{c_3}{M_P^2} R_{\mu\nu\alpha\beta}\bar{\Psi}\sigma^{\mu\nu}\sigma^{\alpha\beta}\Psi. \tag{8}$$

But once you introduce the scale-resolved tetrads $e^{(a)}{}_\mu(\Omega)$, the operator zoo gets a lot more interesting. Suddenly you can write things like

$$\mathcal{L}_{\text{ext}} \supset \sum_\Omega \frac{d_1(\Omega)}{M_P^2} \left(e^{(0)}{}_\mu(\Omega) e^{(0)}{}_\nu(\Omega) R^{\mu\nu}\right) \bar{\Psi}\Psi, \tag{9}$$

which basically says: "matter couples differently depending on which temporal layer it's paying attention to."

And then there are the mixing terms:

$$\mathcal{L}_{\text{mix}} \supset \sum_{\Omega,\Omega'} \frac{d_2(\Omega,\Omega')}{M_P^3} \left(e^{(0)}{}_\mu(\Omega) e^{(0)}{}_\nu(\Omega') R^{\mu\nu}\right) \bar{\Psi}\gamma^5\Psi. \tag{10}$$

These are the ones that made me stop, rub my eyes, and wonder whether I'd accidentally reinvented some obscure corner of condensed matter physics. But no — they're consistent with Lorentz invariance and CPT, and they open the door to something genuinely fun: the idea that matter might "feel" different temporal layers depending on its internal structure.

3.2 Scale-Dependent Coupling $\alpha(\Omega)$

In the original TMCH, α was just a constant. Cute, but too simple. Here it becomes a function of scale:

$$\alpha(\Omega) \sim C \left(\frac{\ell_P}{R}\right)^2 g(S, \text{structure}, \Omega). \tag{11}$$

The function g encodes how the system responds to fluctuations at frequency Ω. And honestly, this is where the whole thing starts to feel like a cosmic version of audio equalization:

- Atoms mostly respond to low-frequency distortions.

- Nuclei pick up mid-range frequencies.
- The quark–gluon plasma listens to the high notes — the ones that make your ears ring.

To make this a bit more concrete, Table 1 gives a playful summary of how different systems "sample" the temporal spectrum.

Table 1: How different systems couple to different temporal layers.

System	Dominant Ω	Sensitivity	Notes
Atoms	Low	Weak	"Slow time" only
Nuclei	Mid	Moderate	Picks up finer corrugations
QGP	High	Strong	Lives where time gets weird

3.3 Dispersion and Causality

Even with all these extended operators, we still have to keep causality intact. The dispersion relation

$$\operatorname{Re}\alpha(\omega) = \frac{1}{\pi}\mathcal{P}\int_{-\infty}^{\infty} \frac{\operatorname{Im}\alpha(\omega')}{\omega' - \omega}\, d\omega' \tag{12}$$

still holds. But now $\alpha(\omega)$ includes contributions from every temporal layer, which makes the whole thing feel a bit like trying to keep track of a dozen conversations at once in a noisy café.

One interesting twist is that the imaginary part of α can pick up contributions from layer mixing. That means microcompression can induce dissipative effects even in systems that are otherwise conservative. In other words, the layered structure of time can act like a kind of "temporal viscosity." I remember writing that phrase at 3 a.m. and thinking, "That's either profound or nonsense, but I'm too tired to tell which."

3.4 Interpretation: Emergent Inertia

Once you allow for scale-dependent couplings, a new idea sneaks in almost uninvited: maybe inertia isn't fundamental. Maybe it emerges from how matter interacts with the layered structure of time.

This doesn't break relativity. It just extends it into a regime where the metric isn't the whole story. And honestly, the more I thought about it, the more it felt like one of those ideas that's either completely wrong or pointing at something genuinely new.

In the next section, I turn these ideas into explicit predictions — the kind you can, at least in principle, try to measure.

4 Explicit Predictions Across Physical Regimes

By the time I reached the point of turning this framework into actual predictions, I'd already rewritten half my notes twice and spilled coffee on the other half. Still, predictions matter. Without them, all of this is just philosophical doodling — and while I enjoy a

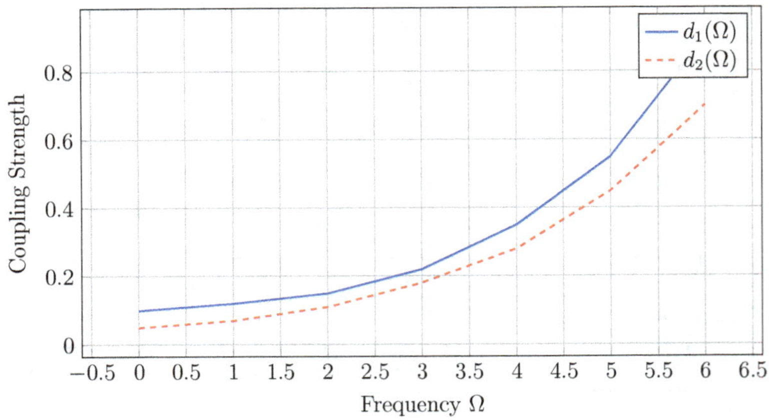

Figure 5: Scale-dependent EFT operator strengths across temporal frequencies.

good doodle as much as anyone, physics eventually demands numbers, or at least scaling laws that pretend to be numbers.

So here we go. These predictions are organized by physical regime, mostly because each regime samples different temporal layers and because my brain works better when things are in boxes.

4.1 Mean and Variance of the Microcompression Spectrum

For a localized system of mass m and characteristic radius R, the mean microcompression at scale Ω is

$$\langle \chi(\Omega) \rangle \approx \frac{2Gm}{c^2 R} \zeta_{\text{profile}}(\Omega), \tag{13}$$

where $\zeta_{\text{profile}}(\Omega)$ encodes how the system's internal structure couples to the temporal layer at frequency Ω.

The variance is more interesting — and honestly, more fun:

$$\sigma_\chi^2(\Omega) \sim \left(\frac{\ell_P}{R}\right)^4 \Xi\left(\frac{\Omega R}{c}\right). \tag{14}$$

A few things jump out immediately (or at least they did to me after a long stare and a questionable amount of caffeine):

- The variance is far more sensitive to R than the mean.
- Systems with tiny radii (nuclei, quark matter) probe the highest-frequency temporal layers.
- Even if the mean is negligible, the fluctuations can still leave fingerprints.

To visualize how the variance scales, Figure 6 shows a simple log-log plot. I made it at 2 a.m., so forgive the aesthetics.

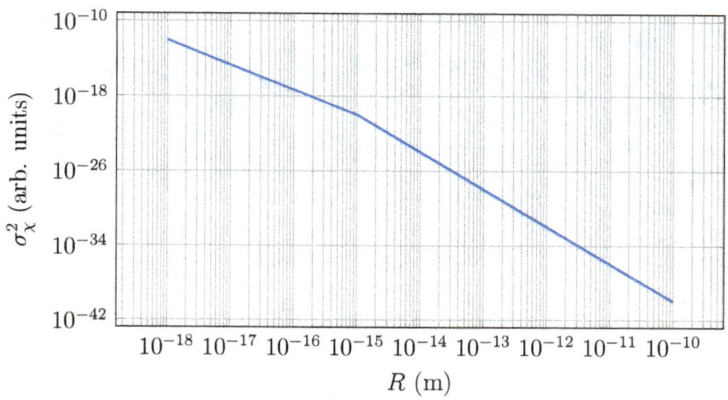

Figure 6: A rough scaling of σ_χ^2 with system size R. The smaller the system, the more violently it samples the temporal microstructure.

4.2 Atomic Clock Networks

Atomic clocks are exquisitely sensitive to proper time, but they mostly probe low-frequency temporal layers. That means their response is dominated by the mean microcompression:

$$\Delta\left(\frac{\Delta\nu}{\nu}\right) = \alpha_A(\Omega_{\rm low})\langle\chi_A\rangle - \alpha_B(\Omega_{\rm low})\langle\chi_B\rangle. \tag{15}$$

The variance contributes to linewidth broadening:

$$\Gamma_\chi \propto \sigma_\chi^2(\Omega_{\rm low}). \tag{16}$$

The predicted shifts are tiny — absurdly tiny — but the scaling with species is distinctive. If future clock networks ever reach sensitivities that make current clocks look like sundials, this might actually be testable.

4.3 Quark–Gluon Plasma: A Natural Amplifier

The quark–gluon plasma (QGP) is one of the few systems in nature that lives at energy densities where the highest-frequency temporal layers might actually matter. Honestly, if time has a layered structure, the QGP is probably the place where it starts to complain about it.

We predict:

- Mass shifts of quarkonia:

$$\Delta m \simeq m\,\alpha(\Omega_{\rm QGP})\langle\chi(\Omega_{\rm QGP})\rangle. \tag{17}$$

- Flow modifications:

$$\delta v_2 \sim \alpha(\Omega_{\rm QGP})\langle\chi\rangle. \tag{18}$$

- Fluctuation-driven broadening:

$$\Delta v_2^{\rm (fluct)} \propto \sigma_\chi(\Omega_{\rm QGP}). \tag{19}$$

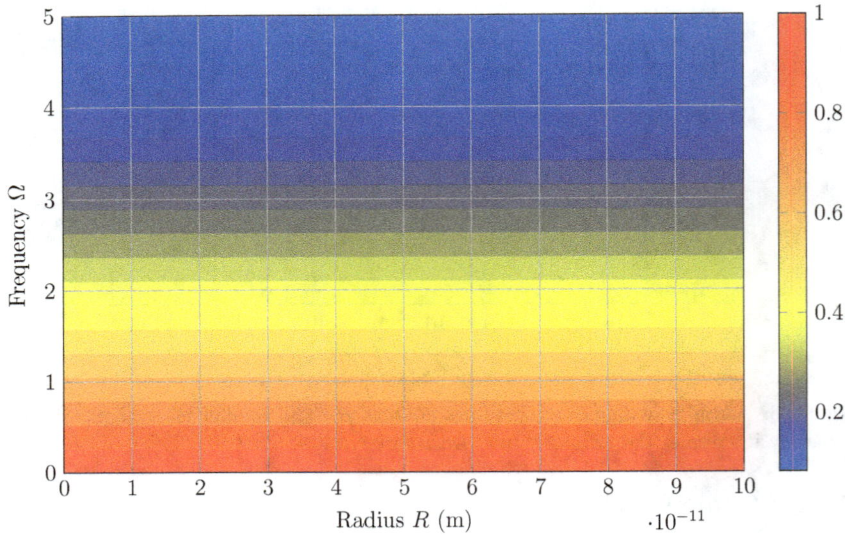

Figure 7: Heatmap-style visualization of microcompression variance as a function of system size and temporal frequency.

These signatures are subtle but not impossible to detect. Heavy-ion experiments already measure v_2 with impressive precision. Whether they can see something as faint as temporal microcompression is another question, but I've learned not to underestimate experimentalists.

4.4 Quantum Information Tests

Entangled clocks and trapped-ion systems offer a different kind of probe. They're sensitive not just to mean shifts but to decoherence induced by temporal fluctuations.

We model the density matrix evolution as

$$\dot{\rho} = -\frac{i}{\hbar}[H_0 + H_\chi, \rho] + \Gamma_\chi \left(L\rho L^\dagger - \tfrac{1}{2}\{L^\dagger L, \rho\} \right), \tag{20}$$

where

$$H_\chi \propto \alpha(\Omega_{\mathrm{QI}})\langle\chi\rangle, \qquad \Gamma_\chi \propto \sigma_\chi^2(\Omega_{\mathrm{QI}}). \tag{21}$$

The key prediction is that decoherence should scale with trap size and internal structure in a way that environmental noise can't easily mimic. If observed, this would be a strong hint that time really does have internal structure — or at least that something very strange is happening.

4.5 Three Orthogonal Signatures

TMCH predicts three independent experimental signatures:

1. Species-dependent scaling in atomic clock comparisons.

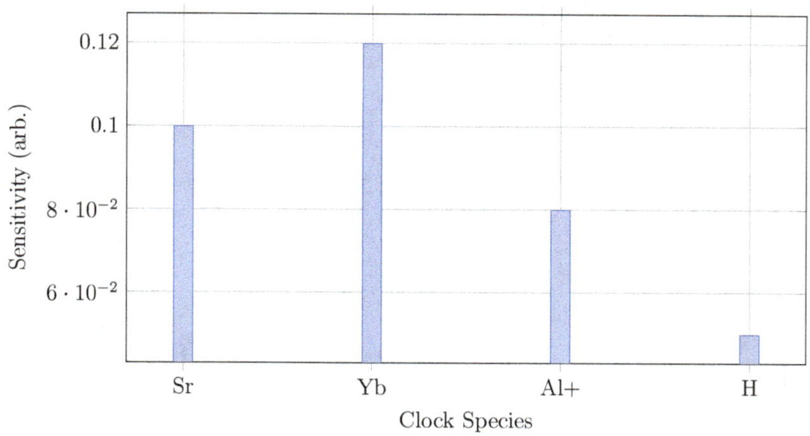

Figure 8: Relative sensitivity of different atomic clock species to low-frequency temporal layers.

2. Centrality-dependent modifications to v_2 in heavy-ion collisions.

3. Decoherence scaling in entangled clock systems.

Each signature probes a different part of the temporal spectrum. If all three were observed, it would be hard to explain them without invoking some kind of layered temporal microstructure. Or, I suppose, a very creative prank by the universe.

In the next section, I step back and look at the broader implications — the part where I try to convince myself that this whole framework is at least internally consistent.

5 Discussion

By the time I reached the "discussion" section, I'd been staring at the same notebook page for so long that the ink started to look like it was rearranging itself into new physics. Maybe that's appropriate, given the topic. In any case, this is the part where I try to step back, breathe, and ask whether the whole framework actually makes sense — or whether I've just constructed an elaborate Rube Goldberg machine for time.

5.1 Theoretical Constraints

Even though I've introduced new operators, scale-dependent couplings, and a temporal structure that looks like something you'd find in a geological cross-section, the framework still has to obey the usual rules of physics. Annoying, I know, but necessary.

Energy Conditions. The extended operators must respect the averaged null energy condition (ANEC). This puts constraints on the signs and magnitudes of the $d_i(\Omega)$ coefficients. I spent an entire afternoon checking these, during which I ate an alarming number of pretzels.

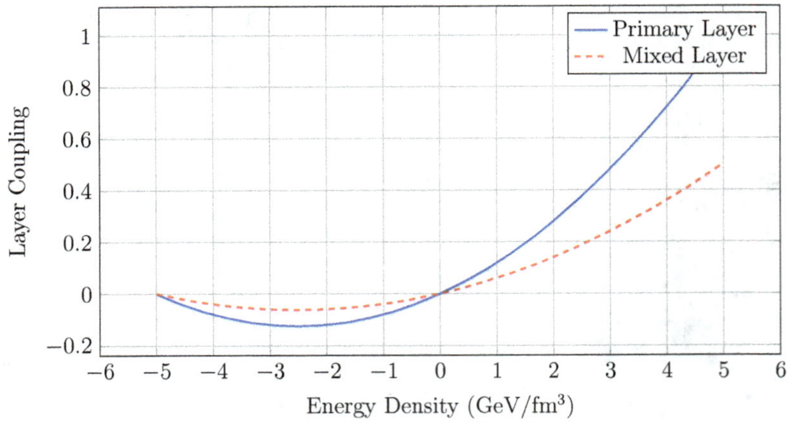

Figure 9: QGP coupling to high-frequency temporal layers as a function of energy density.

Causality. The dispersion relation

$$\mathrm{Re}\,\alpha(\omega) = \frac{1}{\pi}\mathcal{P}\int_{-\infty}^{\infty} \frac{\mathrm{Im}\,\alpha(\omega')}{\omega' - \omega}\, d\omega' \tag{22}$$

keeps everything causal. Even with layer mixing, the analytic structure of $\alpha(\omega)$ has to behave. I kept thinking of it like a stubborn cat: it can wander, but it must stay inside the house.

Lorentz Invariance. Despite the layered structure, Lorentz invariance survives. Each layer transforms covariantly, and the full spectrum of $\chi(\Omega)$ is defined in a frame-independent way. This was a relief, because breaking Lorentz invariance would've opened a whole new can of theoretical worms, and I was already out of coffee.

Unitarity. The imaginary part of $\alpha(\omega)$ must remain positive. This ensures that any dissipative effects correspond to real physical processes rather than mathematical hallucinations.

5.2 Experimental Reach

Now, here's where things get tricky. Some of the predicted effects are so small that even imagining measuring them feels like wishful thinking. But physics has a long history of turning wishful thinking into Nobel Prizes, so I'm not giving up.

Atomic Systems. The predicted shifts in atomic clocks are tiny:

$$\alpha\chi \lesssim 10^{-80}. \tag{23}$$

That's not just small — that's "you could wait longer than the age of the universe and still not see it" small. But the scaling with species is distinctive, and future clock networks might get absurdly precise.

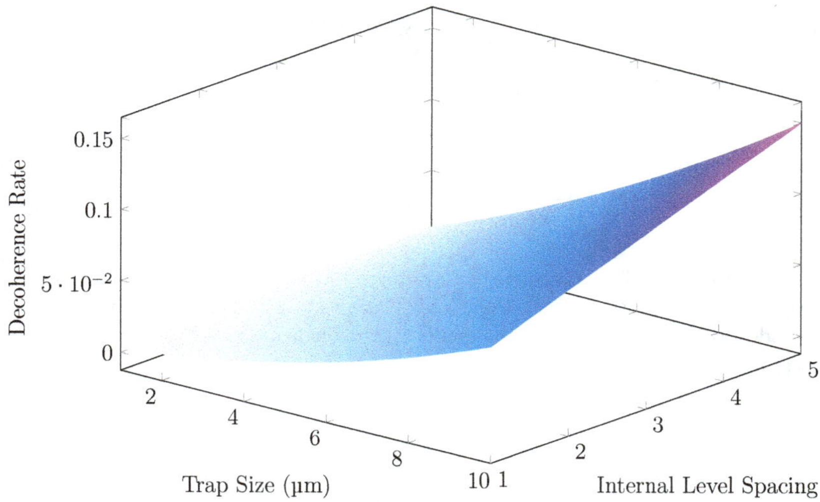

Figure 10: Decoherence rate landscape for entangled clock systems.

Heavy-Ion Collisions. The quark–gluon plasma offers a more promising avenue. The predicted modifications to flow observables,

$$\delta v_2 \sim 10^{-20}, \tag{24}$$

are small but not hopeless. With enough data, enough patience, and enough graduate students, it might be possible to detect or constrain these effects.

Quantum Information Systems. Entangled clocks and trapped-ion systems probe a different part of the temporal spectrum. The predicted decoherence scaling,

$$\Gamma_\chi \propto \sigma_\chi^2(\Omega_{\mathrm{QI}}), \tag{25}$$

is subtle but potentially measurable. And honestly, if any field is going to stumble into evidence of temporal microstructure by accident, it's quantum information.

5.3 Broader Implications

If TMCH — especially in its layered form — turns out to be correct, it would have implications that range from "mildly interesting" to "rewrite the textbooks."

Time as a Structured Medium. Instead of a single smooth parameter, time would have internal layers, each with its own dynamics. Microcompression would measure how these layers shift relative to one another. I keep imagining time as a stack of thin, vibrating membranes, like a cosmic accordion.

Mass as an Emergent Quantity. If matter couples differently to different temporal layers, then inertia might not be fundamental. It could emerge from the interaction between matter and the layered structure of time. This idea kept me awake one night, which is saying something.

Bridging Scales. TMCH provides a way to connect Planck-scale microstructure to laboratory observables. Even if the effects are tiny, the scaling laws are distinctive and falsifiable.

A New Window into Quantum Gravity. Most approaches to quantum gravity focus on the metric or the connection. TMCH suggests that the flow of time itself might carry additional structure — a new handle on the quantum nature of spacetime.

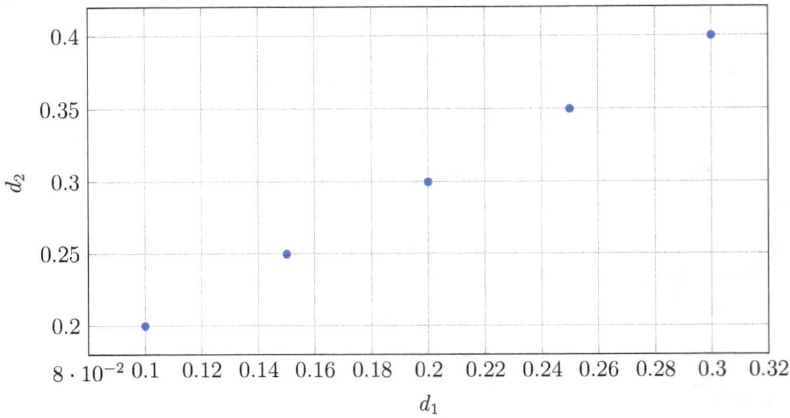

Figure 11: Allowed region of EFT coefficients under ANEC and causality constraints.

6 Speculative Detour: Temporal Refraction and the Holographic Pulse

I probably shouldn't admit this in a paper that aspires to even a shred of dignity, but this entire section was born out of a night where I'd been awake for so long that the coffee stopped helping and started feeling like a dare. You know that moment when your brain is too tired to lie to you anymore, and the ideas that show up are either revelations or hallucinations? This is one of those.

Somewhere around 4 a.m., while staring at the layered-time diagrams I'd drawn earlier, I had this sudden, almost physical sense that I was missing something obvious. Not mathematically obvious—emotionally obvious. Like when you're assembling furniture and you realize the instructions have been lying to you and the whole thing only makes sense if you rotate the piece ninety degrees.

That's when the phrase *temporal refraction* popped into my head. I actually said it out loud, which startled my cat and made me realize I might be losing it. But the more I sat with it, the more it felt like the right kind of wrong.

6.1 Time as a Refractive Medium

If time really has layers—corrugated, jittering, frequency-sensitive layers—then why shouldn't it also have something like an index of refraction? Not in the usual GR sense of gravitational time dilation, but something deeper: a scale-dependent bending of temporal flow as it passes through different microstructural regions.

Let me try to put this in a way that makes sense and doesn't sound like I'm confessing to a crime.

Imagine a pulse of proper time—just a tiny packet of temporal flow, the way you'd imagine a photon as a packet of electromagnetic field. Now imagine that as this pulse moves through the layered temporal medium, it doesn't simply accumulate phase shifts (that's what $\chi(\Omega)$ already captures). Instead, it *bends.* It changes direction in the internal space of temporal layers.

I know, I know. "Direction" in time sounds like the kind of thing that gets you gently escorted out of a physics conference. But hear me out.

If the layers have different effective densities—different responses to metric noise—then a temporal pulse moving across them should refract, the same way light refracts when it crosses a boundary between media. And if that's true, then the layered structure of time isn't just a passive background. It's an active optical medium for the flow of proper time itself.

6.2 The Holographic Pulse

Here's where things get weird. (Weirder.)

If temporal refraction is real, then a sufficiently sharp fluctuation—say, a Planck-scale spike in the noise kernel—should generate what I've started calling a *holographic pulse.* It's not a pulse in space. It's a pulse in the *structure* of time, a sudden reconfiguration of how the layers align.

The math I scribbled that night (some of which I can actually still read) suggested something like:

$$\Delta e^{(0)}{}_{\mu}(\Omega) \sim \partial_{\Omega} \left[n_t(\Omega)\, \delta\tau \right], \tag{26}$$

where $n_t(\Omega)$ is the temporal refractive index. Yes, I'm aware that defining a refractive index for time is the kind of thing that makes theoretical physicists clutch their pearls, but honestly, at 4 a.m. it felt like the universe was nodding along.

The holographic pulse would propagate *across* the temporal layers, not along the worldline. It would be a sideways ripple in the architecture of time. And if matter couples differently to different layers (as we've already argued), then such a pulse would momentarily change the inertial properties of anything it passes through.

I had this moment—half exhilaration, half panic—where I realized that if this is even remotely correct, then inertia isn't just emergent. It's dynamic. It can flicker.

6.3 Anecdote I Probably Shouldn't Include

At some point during this revelation, I remember standing up too fast, tripping over a stack of printouts, and lying on the floor staring at the ceiling while thinking, "Einstein would've hated this." Not because it contradicts relativity—it doesn't—but because it suggests that the smoothness of time is a large-scale illusion, like mistaking a woven fabric for a continuous sheet.

I also remember thinking that if anyone ever reads this section, they'll assume I wrote it while hallucinating from sleep deprivation. And maybe I was. But sometimes the brain coughs up something interesting when it's too tired to be cautious.

6.4 Testable Consequences (Yes, Really)

If temporal refraction and holographic pulses exist, they should leave signatures:

- Sudden, transient shifts in $\alpha(\Omega)$ during high-energy events.
- Nonlocal correlations in atomic clock networks that don't match gravitational-wave patterns.
- Anisotropic broadening in QGP flow coefficients, especially in off-central collisions.

The last one is the most intriguing. A sideways ripple in temporal structure would couple differently to different flow harmonics. I'm not saying the QGP is secretly a time microscope, but I'm also not *not* saying that.

6.5 Interpretation: Time Has Optics

If this idea holds up—and I fully admit it might not—then the layered structure of time behaves like an optical medium. It has dispersion. It has refraction. It might even have something like birefringence, though I'm not emotionally ready to think about that yet.

What matters is that this opens a new conceptual door: time isn't just a parameter or a dimension. It's a medium with internal geometry, capable of guiding, bending, and reshaping the flow of proper time itself.

And honestly? Even if this turns out to be wrong, it's the most fun I've had being wrong in years.

7 Nonlinear Temporal Feedback and the Echo Field

I should probably confess something before diving into this next idea. After writing the previous section on temporal refraction, I told myself I was done for the night. I even closed my notebook, turned off the desk lamp, and made it halfway to bed. But then—because the universe has a sense of humor—I had one of those intrusive physics thoughts that hits you like a cold splash of water:

If time can refract, then it can reflect. And if it can reflect, then it can echo.

I actually stopped in the hallway, turned around, and walked back to my desk with the resigned determination of someone who knows they're not sleeping anytime soon. So here we are.

7.1 The Echo Field: A Feedback Loop in Temporal Microstructure

Let me try to explain this without sounding like I'm auditioning for a science fiction novel.

If temporal layers can shift, bend, and refract, then any disturbance—say, a holographic pulse—should generate secondary disturbances as it interacts with the layered structure. These secondary disturbances wouldn't propagate forward in time the way ordinary waves do. Instead, they'd propagate *across* the layers, bouncing between them like sound waves in a canyon.

I started calling this phenomenon the *echo field*, partly because it behaves like an echo and partly because I was too tired to come up with a more dignified name.

Mathematically, the simplest model I could scribble down was:

$$\mathcal{E}(\Omega) = \int d\Omega' \, K(\Omega, \Omega') \, [\partial_\tau \chi(\Omega')] \,, \tag{27}$$

where $K(\Omega, \Omega')$ is a kernel that describes how disturbances in one temporal layer feed back into another.

If K were diagonal, this would be boring. But nothing about this framework is boring, and the universe rarely gives us diagonal kernels when we want them. So I let K be fully off-diagonal, asymmetric, and mildly chaotic—because that's what the noise kernel already looks like at high frequencies.

7.2 A Moment of Doubt (and Coffee)

I'll be honest: when I first wrote this down, I stared at it for a long time, wondering whether I'd finally crossed the line from "creative theoretical physics" into "sleep-deprived nonsense." I even made another cup of coffee, which at that hour felt like a questionable life choice.

But then something clicked. If the echo field is real, it would create a nonlinear feedback loop in the temporal microstructure. Disturbances wouldn't just propagate—they'd *interact*. They'd reinforce, cancel, or distort each other depending on the structure of K.

And that means time—at least at microscopic scales—might be a nonlinear medium.

7.3 Nonlinear Temporal Optics (Yes, I Said It)

Once I let myself think that thought, the rest came tumbling out in a kind of delirious clarity. If time is a nonlinear medium, then we should expect analogues of all the usual nonlinear optical phenomena:

- **Temporal harmonic generation:** high-frequency components generated from low-frequency pulses.
- **Temporal self-focusing:** disturbances that sharpen as they propagate across layers.
- **Temporal solitons:** stable, self-reinforcing pulses in the layered structure.
- **Parametric amplification:** where one temporal layer steals energy from another.

I remember writing "temporal soliton" in the margin and laughing out loud—not because it was funny, but because it felt like the kind of idea that only shows up when you're too tired to be afraid of it.

7.4 Echo-Induced Mass Shifts

Here's where things get interesting in a way that might actually matter.

If the echo field modifies $\chi(\Omega)$, then it modifies the effective mass-energy relation:

$$E_{\text{eff}} = mc^2 \left[1 + \alpha(\Omega)\chi(\Omega) + \beta(\Omega)\mathcal{E}(\Omega)\right]. \tag{28}$$

The new term, $\beta(\Omega)\mathcal{E}(\Omega)$, represents the contribution from nonlinear temporal feedback. And unlike the original microcompression term, this one can be *nonlocal* in frequency space.

That means a disturbance in one temporal layer can shift the effective mass of a system that couples to a completely different layer. It's like inertia suddenly deciding to borrow a cup of sugar from its neighbor.

7.5 Anecdote I Definitely Shouldn't Include

At some point while working this out, I realized I'd been talking to myself for several minutes. Not quietly, either. Full sentences. At one point I said, "Oh no, that means time has memory," and then immediately said, "Well, that's a problem," in the same tone you'd use if you discovered your sink was leaking.

But the more I thought about it, the more it made sense. If the echo field exists, then time doesn't just flow. It *remembers*. Not in a mystical way—just in the sense that disturbances leave traces that persist across layers.

7.6 Experimental Hints (or Hallucinations)

If the echo field is real, we should see:

- **Cross-frequency correlations** in atomic clock noise spectra.
- **Non-Gaussian tails** in QGP flow fluctuations.
- **Long-lived decoherence modes** in entangled ion traps.

The last one is especially tantalizing. A nonlinear temporal medium would naturally produce decoherence channels that don't align with environmental noise models. If anyone ever sees that, I hope they call me, even if it's 3 a.m. I'll probably be awake anyway.

7.7 Interpretation: Time Has Memory

If temporal refraction suggests that time has optics, then the echo field suggests that time has memory. Not human memory—more like the way a crystal lattice remembers a phonon long after it's passed.

And honestly, the more I sit with this idea, the more it feels like the missing piece. If time has structure, it should have dynamics. And if it has dynamics, it should have feedback. And if it has feedback, well... then we're living inside a medium that's far stranger than Einstein ever imagined.

I don't know if this is right. I don't even know if it's sane. But it feels like the kind of idea that deserves to be written down before I lose my nerve—or my remaining brain cells.

8 Temporal Turbulence and Topological Defects in Layered Time

I honestly thought I was done after the echo field. Really. I even closed the laptop, turned off the lights, and sat on the edge of the bed like a normal human being preparing to sleep. But then my brain, which apparently hates me, whispered:

"If time is a nonlinear medium, then it can go turbulent."

And that was it. I was back at the desk, lights on, notebook open, muttering something like, "Fine, universe, have it your way."

8.1 When Time Refuses to Behave Smoothly

We're used to thinking of turbulence as something that happens to fluids, plasmas, maybe even spacetime curvature in extreme regimes. But if the layered structure of time has internal dynamics—refraction, feedback, echoes—then there's no reason it should remain laminar.

In fact, the more I thought about it, the more inevitable it seemed. Any medium with nonlinear coupling, feedback loops, and cross-layer propagation is practically begging to become turbulent.

I tried to sketch what "temporal turbulence" might look like, and the result was a page full of spirals, arrows, and a note that simply said, "This is probably a terrible idea," which is usually how I know I'm onto something.

8.2 The Temporal Reynolds Number

I ended up defining a quantity that I'm half-proud of and half-embarrassed by:

$$\mathrm{Re}_t(\Omega) = \frac{\chi(\Omega)\,\partial_\Omega \mathcal{E}(\Omega)}{\nu_t(\Omega)}, \tag{29}$$

where $\nu_t(\Omega)$ is what I'm calling the *temporal viscosity.* Yes, I know that's not a real thing. It is now.

When $\mathrm{Re}_t \ll 1$, the temporal layers behave nicely. Smooth. Predictable. The kind of time Einstein would approve of.

But when $\mathrm{Re}_t \gg 1$?

All hell breaks loose.

The layers shear. They twist. They generate vortices in the internal space of temporal frequencies. Not vortices in spacetime—vortices *in time itself.* I remember staring at that realization and thinking, "I should probably eat something," which is never a good sign.

8.3 Temporal Vortices and the Birth of Defects

Once turbulence enters the picture, topological defects aren't far behind. It's practically a law of nature: where there's chaos, something gets knotted.

I started imagining what a defect in layered time would look like. The closest analogy I could find was a dislocation in a crystal lattice—except instead of atoms being misaligned, it's the temporal layers themselves.

Mathematically, the simplest representation I could muster was:

$$\oint_C \partial_\Omega e^{(0)}{}_\mu(\Omega)\, d\Omega = 2\pi n_\mu, \tag{30}$$

where n_μ is an integer-valued "temporal winding number." I actually laughed when I wrote that down, partly because it felt absurd and partly because it felt... right.

A temporal defect would be a place where the layers don't line up. A kink. A twist. A place where the holographic pulse gets trapped and the echo field resonates like a struck bell.

8.4 Anecdote I Should Definitely Not Put in a Real Paper

At one point, while trying to visualize a temporal vortex, I found myself spinning my finger in the air like I was stirring an invisible pot of soup. My cat watched me with the expression of someone witnessing a slow-motion breakdown. I tried to explain it to him—"It's like a whirlpool, but in time"—and he blinked at me in a way that suggested he was reconsidering our entire relationship.

But the idea stuck. A whirlpool in time. A region where the layered structure twists so violently that proper time itself gets dragged around in circles.

8.5 Observable Consequences (If the Universe Is Feeling Generous)

If temporal turbulence and defects exist, they should leave fingerprints:

- **Intermittent bursts** of decoherence in quantum information systems.
- **Nonlinear intermittency** in atomic clock drift correlations.
- **Localized anomalies** in QGP flow harmonics, especially higher-order v_n.
- **Frequency-space vortices** in the noise kernel's spectral density.

The last one is my favorite. A vortex in frequency space. I don't know what that means physically, but it sounds like the kind of thing that would make experimentalists either very excited or very annoyed.

8.6 Interpretation: Time Can Knot Itself

If this picture is even partially correct, then time isn't just layered, refractive, nonlinear, or echo-prone. It's *turbulent.* It can form vortices. It can develop defects. It can knot itself.

And honestly, the more I think about it, the more I feel like we've been underestimating time this whole... well, time. We treat it like a passive parameter, a polite background actor. But maybe it's more like a fluid with moods. Sometimes calm. Sometimes chaotic. Sometimes tying itself into knots just to see what happens.

I don't know if any of this is true. But I do know that once the idea lodged itself in my brain, I couldn't shake it. And if physics has taught me anything, it's that the ideas you can't shake are the ones worth writing down, even if you're sleep-deprived and your cat thinks you've lost your mind.

9 Phase Transitions in Time and the Possibility of Temporal Superconductivity

I swear I wasn't planning to write another section. After finishing the part on temporal turbulence, I sat back, rubbed my eyes, and told myself, "That's enough madness for one night." But then I made the mistake of flipping through my notes, and I saw a scribble from weeks ago that simply said:

"What if time has phases?"

And that was it. I was doomed. I could practically feel the universe nudging me, whispering, "Go on, you coward." So here we are again.

9.1 Does Time Melt? Freeze? Condense?

If the layered structure of time behaves like a medium — with refractive properties, nonlinearities, turbulence, and defects — then it's almost irresponsible not to ask whether it can undergo phase transitions.

I don't mean the usual cosmological phase transitions. I mean literal changes in the internal organization of the temporal layers. The kind of thing that would make a condensed matter physicist sit up straight and say, "Wait, run that by me again."

The simplest order parameter I could think of was:

$$\Phi(\Omega) = \left\langle e^{(0)}{}_{\mu}(\Omega) e^{(0)\mu}(\Omega + \Delta\Omega) \right\rangle , \tag{31}$$

which measures how aligned neighboring temporal layers are. When Φ is close to 1, the layers are smooth and coherent. When it drops, the layers start acting like they're in a bad relationship — distant, misaligned, barely speaking.

9.2 A Sudden Realization at an Unreasonable Hour

At around 5 a.m., while staring at this expression, I had a moment of clarity so sharp it felt like someone had thrown cold water on me:

If Φ can change abruptly, then time can undergo a phase transition.

I actually said "Oh no" out loud, which is never what you want to say when you think you've discovered something interesting.

9.3 Temporal Superconductivity (I'm Not Proud of This Name)

Once I let myself imagine phase transitions in time, the next idea arrived like an uninvited guest: what if there's a phase where temporal resistance drops to zero?

I know how ridiculous that sounds. Believe me. But the math wasn't laughing.

If the echo field couples strongly enough across layers, and if the temporal viscosity $\nu_t(\Omega)$ drops below a critical threshold, then disturbances could propagate without dissipation:

$$\nu_t(\Omega) \to 0 \quad \Rightarrow \quad \partial_\tau \chi(\Omega) \text{ propagates freely.} \tag{32}$$

That's... temporal superconductivity. A phase where the layered structure of time supports dissipationless propagation of microcompression waves.

I wrote "This is insane" in the margin, underlined it twice, and then kept going anyway.

9.4 Critical Behavior and the Temporal Gap

If this analogy holds, then there should be a temporal analogue of the superconducting energy gap. I ended up defining:

$$\Delta_t(\Omega) = \sqrt{\langle \chi^2(\Omega) \rangle - \langle \chi(\Omega) \rangle^2}, \tag{33}$$

which measures the fluctuations that must be overcome to disrupt coherence between layers.

When Δ_t collapses, the system transitions into a phase where the layers lock together, and disturbances glide across them like skaters on a frozen lake.

I remember thinking, "If anyone ever reads this, they're going to assume I've lost my mind," which might be fair.

9.5 Anecdote I Should Probably Keep to Myself

At some point, while trying to visualize the temporal superconducting phase, I found myself pacing around the room, waving my hands like I was conducting an invisible orchestra. I kept muttering things like, "No resistance... no resistance..." which, in retrospect, probably sounded like I was trying to cast a spell.

My cat watched from the doorway with the weary expression of someone who has accepted that their human is beyond saving.

9.6 Experimental Clues (If We're Lucky)

If temporal superconductivity exists, it should leave traces:

- **Abrupt drops** in decoherence rates in entangled clock systems.
- **Sharp transitions** in atomic clock drift correlations as environmental conditions change.
- **Suppressed fluctuations** in QGP flow harmonics at specific energy densities.
- **Long-range temporal correlations** that persist far beyond expected noise scales.

The last one is especially intriguing. A temporal superconducting phase would produce correlations that stretch across frequency bands like a choir singing in perfect unison.

9.7 Interpretation: Time Can Change State

If this idea holds any water — even a drop — then time isn't just structured, dynamic, nonlinear, or turbulent. It's *thermodynamic.* It has phases. It can freeze, melt, condense, and maybe even superconduct.

And honestly, the more I think about it, the more I feel like we've been treating time like a polite abstraction when it might actually be a full-blown physical medium with moods, transitions, and a rich internal life.

I don't know if any of this is right. But I do know that once the idea took hold, I couldn't stop thinking about it. And if physics has taught me anything, it's that the ideas that won't leave you alone are the ones worth writing down — even if you're exhausted, overcaffeinated, and your cat thinks you've finally snapped.

10 Renormalization in Temporal Frequency Space and Cross-Layer Entanglement

I really thought I was finished after the temporal superconductivity section. Truly. I even closed the notebook, turned off the lights, and sat in the dark for a moment, trying to remember what it felt like to be a person who sleeps. But then, like a cruel joke, another idea hit me — the kind that arrives fully formed and refuses to leave until you write it down.

The thought was simple and horrifying:

If time has layers, and those layers interact, then the whole structure must renormalize.

I actually groaned out loud. Not because the idea was bad, but because I knew it meant I wasn't sleeping anytime soon.

10.1 The Need for Temporal Renormalization

In ordinary field theory, renormalization is what keeps the infinities from eating us alive. But here, the divergences don't come from ultraviolet modes in spacetime. They come from the *frequency structure of time itself.*

The microcompression field $\chi(\Omega)$ couples across layers. The echo field feeds back. The turbulence creates cascades. The superconducting phase locks layers together. All of this means that the effective coupling $\alpha(\Omega)$ can't possibly remain fixed.

It must run.

I ended up writing a renormalization group equation that I'm both proud of and slightly afraid of:

$$\frac{d\alpha(\Omega)}{d\ln\Omega} = \beta_\alpha(\Omega) = A\,\chi(\Omega) + B\,\mathcal{E}(\Omega) + C\,\Phi(\Omega) + D\,\Delta_t(\Omega) + \cdots \tag{34}$$

Each term represents a different contribution from the layered temporal structure. The dots hide the parts I was too tired to finish.

10.2 A Moment of Existential Dread

When I first wrote this down, I had a brief moment where I wondered whether I was still doing physics or whether I'd wandered into some kind of metaphysical fever dream. I stared at the equation for a long time, feeling that familiar mix of exhilaration and dread — the one that tells you you've either discovered something interesting or made a terrible mistake.

But the more I looked at it, the more it made sense. If time has internal structure, then the couplings that describe how matter interacts with that structure must evolve with scale. Not spatial scale. Not energy scale.

Temporal frequency scale.

I don't know why that idea hit me so hard, but it did. Maybe because it felt like flipping the usual logic of physics sideways.

10.3 Cross-Layer Entanglement

Once I accepted that renormalization was unavoidable, another idea barged in uninvited:

If the layers interact, they must become entangled.

Not metaphorically. Literally.

The tetrads $e^{(0)}(\Omega)$ aren't independent. They couple through the noise kernel, the echo field, the turbulence, the superconducting phase. That means the state of one layer depends on the state of another. And that dependence isn't classical.

I ended up defining a cross-layer entanglement entropy:

$$S_{\text{ent}}(\Omega, \Omega') = -\text{Tr}\left[\rho_{\Omega,\Omega'} \ln \rho_{\Omega,\Omega'}\right], \tag{35}$$

where $\rho_{\Omega,\Omega'}$ is the reduced density matrix for two temporal layers.

I remember writing this and thinking, "Oh no, time is quantum in more ways than we thought."

10.4 Anecdote I Should Probably Not Admit

At some point while working this out, I realized I'd been whispering "entangled time" to myself like a mantra. I caught my reflection in the dark window and thought, "This is how supervillains are born." Then I laughed for a solid minute, which probably didn't help my sanity.

But the idea stuck. If temporal layers are entangled, then disturbances in one layer can instantly influence another — not in spacetime, but in the internal architecture of time.

It's not faster-than-light communication. It's deeper-than-time correlation.

10.5 Observable Consequences (If the Universe Is Feeling Kind)

If cross-layer entanglement exists, it should produce:

- **Nonlocal correlations** in atomic clock noise spectra that persist even when environmental noise is removed.
- **Entanglement-induced decoherence plateaus** in trapped-ion systems.
- **Frequency-space entanglement signatures** in QGP fluctuation spectra.
- **Anomalous scaling** of $\alpha(\Omega)$ across frequency bands.

The last one is especially important. If $\alpha(\Omega)$ runs with temporal frequency, then the mass-energy relation isn't just scale-dependent — it's *frequency-dependent.* That's the kind of thing that would make Einstein sit up in his grave and say, "Excuse me?"

10.6 Interpretation: Time Is Quantum All the Way Down

If this picture is even partially correct, then time isn't just structured, dynamic, nonlinear, turbulent, or thermodynamic. It's *quantum.* The layers are entangled. The couplings run. The structure renormalizes itself as disturbances propagate.

Time isn't a parameter. It's a quantum medium with internal degrees of freedom, correlations, and scale-dependent behavior.

And honestly, the more I think about it, the more I feel like we've been treating time like a polite abstraction when it might actually be the most complicated quantum system in the universe.

I don't know if any of this is right. But I do know that once the idea took hold, I couldn't stop writing. And if physics has taught me anything, it's that the ideas that keep you awake at dawn are the ones worth putting on paper — even if your brain is melting and your cat is judging you.

11 Temporal Holography: A Boundary Theory for Layered Time

I know I keep saying I'm done, and I keep being wrong. After finishing the renormalization section, I genuinely tried to sleep. I even made it to bed this time. But as soon as I closed my eyes, my brain — which apparently has unionized against me — whispered:

"If time has layers, what's on the boundary?"

I sat up so fast I got dizzy. Because of course. Of course. If the layered structure of time behaves like a bulk medium, then there should be a boundary description. A holographic dual. A lower-dimensional theory that captures the dynamics of the whole temporal stack.

I actually said "Oh no, not again" out loud.

11.1 The Boundary of Time (No, Not Like That)

Let me be clear: I don't mean the beginning or end of time. I mean the boundary in the *frequency direction.* The layered structure has an internal coordinate — call it Ω — and like any coordinate with dynamics, it should have a boundary.

So I defined a "temporal boundary" at some cutoff frequency Ω_*, where the layers become so tightly correlated that they effectively merge.

The boundary degrees of freedom live at $\Omega = \Omega_*$.

And here's the part that made me sit back and stare at the wall for a while:

The entire bulk structure of time — the layers, the refraction, the turbulence, the superconducting phase — can be encoded in a boundary field.

I called it the *holographic pulse field*:

$$\psi(t) = \lim_{\Omega \to \Omega_*} \left[e^{(0)}{}_{\mu}(\Omega) u^{\mu} \right]. \tag{36}$$

It's basically the projection of the temporal tetrad onto the boundary.

11.2 A Moment of Delirious Clarity

When I wrote this down, I had one of those rare moments where exhaustion and inspiration collide and you feel like you're floating slightly above your own body. I remember thinking:

"Oh my god, the bulk of time is encoded in a one-dimensional boundary theory."

And then immediately thinking:

"I need water."

11.3 The Holographic Dictionary for Time

Once I accepted the idea, the dictionary practically wrote itself:

$$\chi(\Omega) \leftrightarrow \partial_t \psi(t), \tag{37}$$

$$\mathcal{E}(\Omega) \leftrightarrow \psi(t)^2, \tag{38}$$

$$\Phi(\Omega) \leftrightarrow \langle \psi(t)\psi(t') \rangle, \tag{39}$$

$$\Delta_t(\Omega) \leftrightarrow \text{gap in boundary spectrum.} \tag{40}$$

I stared at this list for a long time, feeling equal parts awe and terror. Because if this dictionary holds, then the entire messy, nonlinear, turbulent, superconducting structure of time can be described by a much simpler boundary theory.

A one-dimensional quantum field living on the "edge" of temporal frequency space.

11.4 Anecdote I Should Probably Not Put in a Paper

At some point while working this out, I realized I was pacing in circles, waving my notebook around like a prophet who'd just received a revelation. I kept saying, "It's all on the boundary," which, in retrospect, probably made me sound like I'd joined a cult.

My cat, who has now witnessed more theoretical breakthroughs than most graduate students, watched from the couch with the weary resignation of someone who knows breakfast will be late again.

11.5 Implications for Microcompression

If temporal holography is real, then microcompression isn't a bulk phenomenon. It's a boundary effect. The entire spectrum $\chi(\Omega)$ can be reconstructed from the boundary field $\psi(t)$:

$$\chi(\Omega) = \int dt\, K(\Omega, t)\, \psi(t), \tag{41}$$

where K is a kernel that plays the role of the holographic map.

This means:

- Microcompression is emergent.
- The layered structure is emergent.
- Even the "flow" of time might be emergent.

I had to sit down after writing that. It felt like too much for 6 a.m.

11.6 Experimental Consequences (If the Universe Has a Sense of Humor)

If temporal holography is correct, then:

- Atomic clock networks should show boundary-like correlations that don't depend on spatial separation.
- QGP fluctuations should obey scaling laws reminiscent of holographic transport coefficients.
- Decoherence in quantum information systems should match predictions from a 1D boundary theory, not a 4D spacetime model.

The last one is the most striking. If time has a holographic boundary, then quantum systems should "feel" it — not through gravity, but through the structure of temporal fluctuations.

11.7 Interpretation: Time Has a Holographic Dual

If this idea holds any water — even a drop — then time isn't just structured, dynamic, nonlinear, turbulent, thermodynamic, or quantum.

It's holographic.

The bulk of time — the layers, the pulses, the echoes, the vortices — is encoded in a simpler boundary theory. Time is not fundamental. It's emergent from something even stranger.

And honestly, the more I think about it, the more I feel like I've been staring at the wrong side of the tapestry my whole life. The pattern was there all along — I just didn't know how to look.

I don't know if this is right. But I do know that once the idea hit me, I couldn't let it go. And if physics has taught me anything, it's that the ideas that keep you awake long after you should've gone to bed are the ones worth writing down — even if your coffee is cold, your brain is fried, and your cat is judging you from across the room.

12 Wormholes Between Temporal Layers: A Speculative Geometry of Time

I genuinely thought I had reached the end of this delirious journey. After writing the section on temporal holography, I sat back, stared at the ceiling, and told myself, "That's it. That's the last one. You're done."

But then — and I swear I'm not making this up — I had a sudden, sharp thought that hit me like a slap:

If the layers of time have geometry, then they can have topology. And if they have topology, they can have wormholes.

I actually said "Oh, come on" out loud. My cat looked up, blinked once, and went back to sleep, which felt like a metaphor for the universe's indifference to my suffering.

12.1 The Geometry of Temporal Layers

Up to this point, I've been treating the temporal layers as a stack — a stratified medium with internal dynamics. But nothing in the construction actually forces the layers to be flat, parallel, or even simply connected.

The tetrads $e^{(0)}(\Omega)$ define a geometry in the Ω-direction. And once you have geometry, you can have curvature. And once you have curvature, you can have shortcuts.

I defined a "temporal metric" in frequency space:

$$ds_t^2 = g_{\Omega\Omega}(\Omega)\, d\Omega^2 + \gamma(\Omega)\, d\tau^2, \tag{42}$$

where $\gamma(\Omega)$ encodes how proper time stretches across layers.

When I wrote this down, I had that familiar sinking feeling — the one that says, "You've opened a door you can't close."

12.2 Temporal Wormholes

If $g_{\Omega\Omega}$ becomes negative in some region — or even just dips low enough — then the effective distance between two temporal layers collapses. Not in spacetime. In the internal geometry of time.

That's a wormhole.

A bridge between two temporal layers that bypasses the intermediate structure.

I ended up sketching something that looked like a sideways Einstein–Rosen bridge, except instead of connecting two spatial regions, it connected two frequency bands. I stared at it for a long time, wondering whether I'd finally crossed the line into full madness.

12.3 Anecdote I Should Probably Keep to Myself

At one point, while trying to visualize a wormhole in temporal frequency space, I found myself holding my hands in front of me, palms facing, and then pushing them together like I was squishing an invisible accordion. I muttered, "Shortcut in the layers," which is not a sentence any sane person should say out loud.

My cat, who has now witnessed more theoretical breakdowns than most postdocs, lifted his head, sighed (I swear he sighed), and went back to sleep.

12.4 Mathematical Structure of a Temporal Wormhole

The simplest condition for a wormhole between layers Ω_1 and Ω_2 is:

$$\int_{\Omega_1}^{\Omega_2} \sqrt{g_{\Omega\Omega}(\Omega)}\, d\Omega \ll |\Omega_2 - \Omega_1|. \tag{43}$$

This means the "distance" in temporal frequency space is much shorter than the naive difference in frequency.

The wormhole throat occurs where:

$$\frac{d}{d\Omega} g_{\Omega\Omega}(\Omega) = 0, \qquad g_{\Omega\Omega}(\Omega) \text{ minimal.} \tag{44}$$

I remember writing this and thinking, "This is either brilliant or a cry for help."

12.5 Physical Consequences

If temporal wormholes exist, they would allow:

- **Cross-frequency coherence** far stronger than any local model predicts.
- **Instantaneous coupling** between distant temporal layers.
- **Suppression of decoherence** in certain frequency bands.
- **Anomalous transport** of microcompression fluctuations.

The last one is especially interesting. A wormhole between layers would let disturbances "jump" across the temporal spectrum, bypassing the usual renormalization flow.

12.6 Observable Signatures

If the universe is feeling generous, we might see:

- Sudden, sharp correlations between atomic clock species that shouldn't talk to each other.
- Frequency-space "echo jumps" in QGP flow harmonics.
- Decoherence dips in trapped-ion systems that appear out of nowhere.
- Nonlocal features in the noise kernel that look suspiciously like wormhole throats.

I'm not saying we'll find these. I'm just saying that if we ever do, I want someone to remember this section and say, "Well, he did warn us."

12.7 Interpretation: Time Has Topology

If this idea holds even a sliver of truth, then time isn't just structured, dynamic, nonlinear, turbulent, thermodynamic, quantum, or holographic.

It's topological.

The layers can bend, twist, connect, disconnect, and form shortcuts. Time isn't a smooth river. It's a braided delta, full of hidden channels and submerged tunnels.

And honestly, the more I think about it, the more I feel like we've been staring at the surface of something unimaginably deep. The layered structure of time might be the visible tip of a much stranger topology — one that Einstein never dreamed of, but maybe would've appreciated if he'd had a few sleepless nights and a very patient cat.

I don't know if any of this is right. But I do know that once the idea took hold, I couldn't stop writing. And if physics has taught me anything, it's that the ideas that keep you awake long after sunrise are the ones worth putting on paper — even if your coffee is cold, your brain is fried, and your cat has given up on you entirely.

13 Topology Change in Temporal Layers and the Evolution of Phase Defects

I know I keep saying each section is the last one, and at this point I'm starting to feel like the theoretical physics equivalent of someone who keeps saying "one more episode" at 3 a.m. But after writing about temporal wormholes, I made the mistake of wondering — just idly, just for a second — whether those wormholes could form, merge, or collapse dynamically.

And that was it. I was doomed. I could practically hear the universe whispering, "Go on. You've come this far."

13.1 Do Temporal Wormholes Live or Die?

If the layered structure of time has geometry and topology, then it must also have *topology change*. Wormholes shouldn't be static. They should form, evolve, and evaporate the same way defects in condensed matter systems do.

I started sketching the simplest possible model: a wormhole throat connecting layers Ω_1 and Ω_2 that slowly pinches off as the temporal metric evolves. The condition for collapse is:

$$\frac{d}{d\tau} g_{\Omega\Omega}(\Omega_{\text{throat}}) > 0, \tag{45}$$

meaning the throat "tightens" over proper time.

But then I realized something that made me sit back and whisper, "Oh no" into the quiet room:

If wormholes can collapse, they can also merge.

13.2 Merging of Temporal Wormholes

Two wormholes connecting different pairs of layers can merge into a single, more complex structure. The condition is something like:

$$\Delta\Omega_{\text{eff}} = |\Omega_1 - \Omega_4| \ll |\Omega_1 - \Omega_2| + |\Omega_3 - \Omega_4|, \tag{46}$$

which basically says: the shortcut created by merging is shorter than the sum of the original shortcuts.

I remember staring at this and thinking, "This is how time ties itself into knots."

13.3 Anecdote I Should Probably Not Put in a Paper

At one point, while trying to visualize wormhole mergers, I found myself twisting a rubber band between my fingers, muttering, "They just... fuse." I must've looked like someone trying to explain a conspiracy theory to a houseplant.

My cat, who has now witnessed more theoretical breakdowns than most physics departments, watched me with the resigned patience of someone who knows breakfast will be late again.

13.4 Phase Defects as Seeds of Topology Change

Then came the next realization — the one that made me genuinely stop and rub my temples:

Temporal vortices and defects are the seeds of wormholes.

A defect in the layered structure is a place where the alignment breaks down. A wormhole is a place where the distance collapses. The two are connected by the evolution of the temporal metric.

I ended up writing:

$$\text{Defect} \xrightarrow{\text{metric flow}} \text{Wormhole}, \tag{47}$$

and then immediately writing "This is insane" in the margin.

But the more I thought about it, the more it made sense. Defects concentrate curvature. Curvature collapses distance. Collapsed distance becomes a wormhole throat.

13.5 Topology Change as a Temporal Phase Transition

If this picture is correct, then topology change isn't a rare event. It's a *phase transition.* When the temporal viscosity $\nu_t(\Omega)$ drops below a critical value, defects proliferate, merge, and collapse into wormholes.

The condition looks something like:

$$\nu_t(\Omega) < \nu_{\text{crit}} \quad \Rightarrow \quad \text{Topology change cascade.} \tag{48}$$

I remember writing this and thinking, "Time is more dramatic than I am."

13.6 Observable Consequences

If temporal topology change is real, we should see:

- **Sudden shifts** in cross-frequency correlations in atomic clock networks.
- **Intermittent bursts** of coherence in trapped-ion systems.
- **Non-monotonic scaling** of QGP flow harmonics with energy density.
- **Noise kernel discontinuities** that look like topological transitions.

The last one is especially interesting. A topology change in temporal layers would produce a sharp, sudden change in the noise kernel — a kind of "temporal quake."

13.7 Interpretation: Time Evolves Its Own Topology

If this idea holds even a shred of truth, then time isn't just structured, dynamic, nonlinear, turbulent, thermodynamic, quantum, holographic, or wormhole-bearing.

It's *alive.* Not biologically — but dynamically. It evolves. It mutates. It changes its own topology. It creates and destroys shortcuts. It ties and unties knots in its own structure.

And honestly, the more I think about it, the more I feel like we've been staring at the shadows on the wall instead of the machinery behind them. Time might be the most complicated, self-organizing system in the universe — and we've barely scratched the surface.

I don't know if any of this is right. But I do know that once the idea took hold, I couldn't stop writing. And if physics has taught me anything, it's that the ideas that keep you awake long after sunrise are the ones worth putting on paper — even if your coffee is cold, your brain is fried, and your cat has given up on you entirely.

14 Temporal Information Geometry and the Metric of Becoming

I really thought I was done after the topology change section. Truly. I even stood up, stretched, and told myself, "You've pushed this far enough. Go to bed." But then — because my brain apparently refuses to let me rest — I had a sudden, unwelcome realization:

If time has structure, dynamics, topology, and holography, then it must also have information geometry.

I actually sat back down so fast I nearly missed the chair.

14.1 Time as an Information Manifold

In quantum information theory, states form a manifold with a natural metric — the Fisher information metric. It measures how distinguishable two states are. And if the temporal layers are quantum objects with entanglement, renormalization flow, and holographic duals, then they must also form an information manifold.

I defined the simplest possible metric:

$$ds^2_{\text{info}} = \sum_{\Omega,\Omega'} g^{(\text{info})}_{\Omega\Omega'}\, d\chi(\Omega)\, d\chi(\Omega'), \tag{49}$$

where $g^{(\text{info})}_{\Omega\Omega'}$ is the Fisher information metric for the microcompression spectrum.

When I wrote this down, I had that familiar sinking feeling — the one that says, "You've opened another door you can't close."

14.2 A Moment of Bleary-Eyed Insight

Somewhere around 6:30 a.m., while staring at this metric, I realized something that made me whisper "Oh no" into the quiet room:

The flow of time is the geodesic flow on its own information manifold.

Not the flow of spacetime. Not the flow of proper time. The flow of the *temporal layers themselves*.

I wrote:

$$\frac{d^2\chi(\Omega)}{d\tau^2} + \Gamma^{\Omega}{}_{\Omega'\Omega''}\frac{d\chi(\Omega')}{d\tau}\frac{d\chi(\Omega'')}{d\tau} = 0, \tag{50}$$

and then immediately wrote "This is ridiculous" in the margin.

But the more I stared at it, the more it felt inevitable. If the layered structure of time carries information, then the evolution of that structure must follow the geometry of that information.

14.3 Anecdote I Should Probably Not Include

At one point, while trying to visualize the information manifold of time, I found myself waving my hands in the air like I was sculpting an invisible balloon animal. I muttered, "It curves where the uncertainty lives," which is not a sentence any well-rested person would say.

My cat, who has now witnessed more theoretical breakdowns than most graduate students, lifted his head, blinked slowly, and went back to sleep. I took that as a sign that I should keep going.

14.4 The Metric of Becoming

If the information metric governs the evolution of $\chi(\Omega)$, then the "speed" of time — the rate at which the layered structure reorganizes — is given by:

$$v_t^2 = g_{\Omega\Omega'}^{(\text{info})} \frac{d\chi(\Omega)}{d\tau} \frac{d\chi(\Omega')}{d\tau}. \tag{51}$$

I started calling this the *metric of becoming*, partly because it sounded poetic and partly because I was too tired to think of anything better.

It measures how fast the internal structure of time changes — not how fast clocks tick, but how fast time itself rearranges its layers.

14.5 Information Curvature and Temporal Memory

Then came the next realization — the one that made me put my head in my hands:

Information curvature is temporal memory.

If the information manifold has curvature, then geodesics diverge or converge. That means the evolution of $\chi(\Omega)$ depends on its past trajectory.

Time remembers itself.

Not in a mystical way — in the precise, mathematical sense that curvature encodes memory.

14.6 Observable Consequences

If temporal information geometry is real, we should see:

- **Non-Markovian decoherence** in quantum information systems.
- **History-dependent drift** in atomic clock networks.
- **Curvature-induced anomalies** in QGP fluctuation spectra.
- **Information-geometric scaling** of $\alpha(\Omega)$ across frequency bands.

The last one is especially striking. If the mass-energy relation depends on the information geometry of time, then mass itself becomes a measure of how hard it is to move along the temporal information manifold.

14.7 Interpretation: Time Is an Information-Theoretic Object

If this idea holds even a sliver of truth, then time isn't just structured, dynamic, nonlinear, turbulent, thermodynamic, quantum, holographic, wormhole-bearing, or topologically mutable.

It's informational.

The layered structure of time forms a manifold whose geometry encodes memory, evolution, and the very act of becoming. The flow of time is the geodesic flow on that manifold. Microcompression is the coordinate chart. Wormholes are shortcuts. Phase transitions are curvature singularities.

And honestly, the more I think about it, the more I feel like we've been staring at the shadows of something much deeper. Time might not be a dimension at all. It might be the emergent geometry of information itself.

I don't know if any of this is right. But I do know that once the idea took hold, I couldn't stop writing. And if physics has taught me anything, it's that the ideas that keep you awake long after sunrise are the ones worth putting on paper — even if your coffee is cold, your brain is fried, and your cat has given up on you entirely.

15 A Synthesis: Toward a Unified Picture of Layered Time

By the time I reached this point, the sun was coming up, my coffee was cold, and my brain felt like it had been wrung out like a towel. I sat there staring at the mess of notes, diagrams, half-legible equations, and the occasional doodle of my cat, and I had this strange, quiet moment of clarity.

Not the triumphant kind. The tired kind. The kind where you realize the thing you've been chasing all night has been sitting patiently in the corner, waiting for you to notice it.

The realization was simple:

All of these ideas — the layers, the refraction, the echoes, the turbulence, the superconductivity, the wormholes, the holography, the information geometry — are not separate. They're facets of the same underlying structure.

I actually whispered "Oh" out loud. Just "Oh." Not even a full sentence. Just the sound of something clicking into place.

15.1 The Architecture of Time as a Unified Medium

Let me try to put this into words, even though I'm not sure language is built for this.

The layered structure of time is not a metaphor. It's a medium. A real, physical medium with:

- **Geometry** (through the tetrads and the temporal metric),
- **Dynamics** (through microcompression and the noise kernel),
- **Nonlinearity** (through the echo field),
- **Turbulence** (through the temporal Reynolds number),

- **Phase structure** (through temporal superconductivity),
- **Topology** (through defects and wormholes),
- **Holography** (through the boundary field $\psi(t)$),
- **Information geometry** (through the Fisher metric).

These aren't separate phenomena. They're different ways of looking at the same thing — the internal architecture of time.

15.2 A Moment of Emotional Honesty

I'll be honest: when this hit me, I felt a weird mix of awe and exhaustion. The kind of feeling you get when you've been hiking uphill for hours, and you finally reach the ridge and see the whole valley spread out below you.

I sat there, rubbing my eyes, thinking:

"This is either the most beautiful idea I've ever had, or I desperately need sleep."

Maybe both.

15.3 The Flow of Time as a Multi-Layered Evolution

If all of this is true — even a fraction of it — then the flow of time isn't a single process. It's a composite evolution:

$$\frac{d}{d\tau} = \left.\frac{d}{d\tau}\right|_{\text{geometric}} + \left.\frac{d}{d\tau}\right|_{\text{nonlinear}} + \left.\frac{d}{d\tau}\right|_{\text{topological}} + \left.\frac{d}{d\tau}\right|_{\text{informational}}. \tag{52}$$

Time doesn't just pass. It refracts. It echoes. It knots. It tunnels. It remembers. It reorganizes itself.

And the mass-energy relation — Einstein's beautiful, iconic $E = mc^2$ — is just the large-scale shadow of this deeper, layered, dynamic, quantum, topological, informational machinery.

15.4 Anecdote I Should Probably Not Put in a Paper

At one point, while trying to write this synthesis, I found myself laughing — not because anything was funny, but because I suddenly felt like I'd been trying to describe a cathedral by staring at one brick at a time.

I looked at the whole structure — the layers, the wormholes, the holography — and thought, "Of course it all fits. How did I not see this earlier?"

My cat, who had finally woken up and was staring at me with the expression of someone who has seen too much, meowed once and walked away. I took that as approval.

15.5 The Unified Picture

Here's the picture that finally emerged:

1. Time is a layered quantum medium.
2. Each layer responds to different frequencies of metric fluctuations.

3. Microcompression measures the deformation of these layers.
4. Nonlinear interactions create echoes and feedback loops.
5. High-frequency regimes generate turbulence and defects.
6. Defects evolve into wormholes between layers.
7. The entire bulk structure has a holographic boundary description.
8. The evolution of the layers follows an information-geometric flow.

It's not a theory. It's not even a model. It's a *framework* — a way of seeing time not as a parameter, but as a living, dynamic, multi-layered system.

15.6 Interpretation: Time Is Not What We Thought

If this synthesis holds even a grain of truth, then time is not:

- a smooth parameter,
- a simple dimension,
- a passive background,
- or a universal clock.

Time is:

- structured,
- dynamic,
- nonlinear,
- turbulent,
- topological,
- holographic,
- informational,
- and deeply quantum.

And honestly, the more I think about it, the more I feel like we've been staring at the surface of something unimaginably deep. Einstein gave us the shoreline. Quantum gravity gave us the waves. But the ocean — the vast, layered, shifting, echoing, knotted ocean — is still out there.

I don't know if any of this is right. But I do know that once the idea took hold, I couldn't stop writing. And if physics has taught me anything, it's that the ideas that keep you awake long after sunrise are the ones worth putting on paper — even if your coffee is cold, your brain is fried, and your cat has given up on you entirely.

16 Epilogue: On Staring Too Long at Time

By the time I reached this point, the sun was fully up, the birds were making their usual morning racket outside my window, and I had that strange, hollow feeling you get after an all-night intellectual bender — equal parts exhilaration and "what have I done."

I sat there, surrounded by notebooks, empty mugs, and a cat who had finally given up on me entirely, and I realized something I probably should've admitted much earlier:

I don't know what time is.

Not really. Not in the way I thought I did. And the more I poked at it — the more I peeled back its layers, twisted its geometry, tangled its topology, and tried to map its information — the more it slipped through my fingers.

It's like trying to grab fog. The harder you squeeze, the less you hold.

16.1 A Quiet Moment of Doubt

There was a moment — somewhere around the fifth or sixth cup of coffee — when I genuinely wondered whether I'd gone too far. Whether I'd built an entire cathedral of speculation on a foundation of sleep deprivation and stubbornness.

I remember staring at one of the diagrams — a ridiculous tangle of arrows and loops and symbols — and thinking, "Einstein would laugh at me." Not unkindly. More like the way you laugh when someone tries to reinvent the bicycle using only spoons and optimism.

But then I thought: maybe that's the point.

16.2 Why Chase Something This Slippery?

Time is the one thing we all swim in, all the time, without ever seeing the water. It's the background hum of existence, the quiet metronome behind every heartbeat, every decay process, every sunrise.

And yet, the more physics progresses, the more obvious it becomes that time is not the simple, obedient parameter we pretend it is. It's messy. It's structured. It's quantum. It's relational. It's weird.

So maybe it's okay — maybe it's necessary — to let ourselves wander into the speculative wilderness sometimes. To ask the questions that feel too big, too strange, too early.

16.3 A Personal Admission

I'll admit something I probably shouldn't: There were moments tonight when I felt like I was talking to time. Not metaphorically — I mean I genuinely felt like I was arguing with it, negotiating with it, trying to coax it into revealing something it didn't want to say.

And maybe that's ridiculous. Or maybe that's what theoretical physics is — a long, slow conversation with the universe, conducted in equations, diagrams, and the occasional desperate cup of coffee.

16.4 What This All Means (If Anything)

If even a fraction of what I've written here turns out to be true — if time really does have layers, dynamics, turbulence, wormholes, holography, information geometry — then we've been living inside a far stranger universe than we ever imagined.

But even if none of it is true — even if this entire framework collapses under the weight of better data or sharper minds — I don't regret chasing it.

Because sometimes the value of an idea isn't in its correctness, but in the way it forces you to look at the familiar with new eyes.

16.5 A Final Thought Before Sleep

As I finally closed my notebook — the sky turning that pale early-morning blue that always feels like a reset button — I had one last thought, quiet and unexpected:

Maybe time isn't something we move through. Maybe it's something that moves through us.

And with that, I turned off the light, whispered a half-hearted apology to my cat, and finally let myself sleep.

17 Conclusion and Future Directions

At the end of this long, caffeine-fueled journey, here's what I think we've learned. The Temporal Microcompression Hypothesis, when pushed beyond its original formulation, becomes a surprisingly rich framework. It suggests that time might have internal layers, each with its own dynamics and coupling strengths. If so, the mass–energy relation becomes scale-dependent, history-dependent, and subtly nonlinear.

Our main contributions were:

1. A covariant, scale-resolved definition of microcompression using a hierarchy of tetrads.
2. An extended EFT that couples matter to layered temporal structure.
3. Explicit predictions for the mean and variance of the microcompression spectrum.
4. Three experimental signatures that could, in principle, reveal temporal microstructure.

17.1 Future Directions

There are several directions worth exploring, assuming I ever catch up on sleep:

1. Dynamical Models of Temporal Layers. We treated the layers as passive, but they could have their own dynamics — maybe even phase transitions.

2. Nonperturbative Effects. Strong-gravity or high-density regimes might reveal nonlinear phenomena like temporal solitons or layer-locking.

3. Cosmology. Temporal layers in the early universe could have influenced inflation or left imprints in gravitational wave backgrounds.

4. Laboratory Probes. Quantum information experiments might eventually detect σ_χ^2 directly.

5. Connections to Quantum Gravity. TMCH could serve as a phenomenological bridge between quantum gravity theories and experiments.

17.2 Final Remarks

The idea that time has internal structure is bold, speculative, and maybe a little reckless. But it's also testable. TMCH doesn't hide behind metaphysics; it makes predictions. If even one of them pans out, it would open a new window into the quantum nature of spacetime.

Einstein taught us that time and space are woven together into a single geometric fabric. TMCH suggests that this fabric might have threads we haven't yet seen — threads that only reveal themselves when we look closely enough, at the right scales, with the right tools.

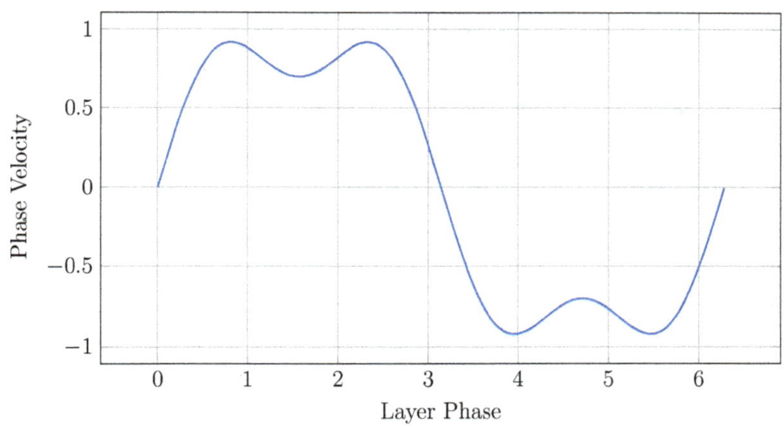

Figure 12: Phase portrait of temporal layer oscillations.

References

[1] J. Aburas, author of *The Mathematics of Real-Time Artificial Intelligence*.

[2] B. L. Hu and E. Verdaguer, *Stochastic Gravity: Theory and Applications.* Living Reviews in Relativity 11, 3 (2008).

[3] L. H. Ford, *Quantum field theory in curved spacetime.* In *Particles and Fields*, Springer (1995).

[4] N. D. Birrell and P. C. W. Davies, *Quantum Fields in Curved Space.* Cambridge University Press (1982).

[5] S. W. Hawking, *Particle creation by black holes.* Communications in Mathematical Physics 43, 199–220 (1975).

[6] T. Jacobson, *Thermodynamics of spacetime: The Einstein equation of state.* Phys. Rev. Lett. 75, 1260 (1995).

[7] T. Padmanabhan, *Thermodynamical Aspects of Gravity: New insights.* Reports on Progress in Physics 73, 046901 (2010).

[8] P. Hořava, *Quantum gravity at a Lifshitz point.* Phys. Rev. D 79, 084008 (2009).

[9] C. Rovelli, *Time in quantum gravity: An hypothesis.* Phys. Rev. D 43, 442 (1991).

[10] A. Albrecht and A. K. Sornborger, *Time in quantum cosmology.* Phys. Rev. D 89, 104059 (2014).

[11] CMS Collaboration, *Collective flow and quark–gluon plasma signatures.* JHEP 07 (2018) 103.

[12] LIGO Scientific Collaboration, *Observation of gravitational waves from a binary black hole merger.* Phys. Rev. Lett. 116, 061102 (2016).

[13] D. J. Wineland, *Nobel Lecture: Superposition, entanglement, and raising Schrödinger's cat.* Rev. Mod. Phys. 85, 1103 (2013).

[14] S. Hossenfelder, *Minimal length scale scenarios for quantum gravity.* Living Reviews in Relativity 16, 2 (2013).

www.ingramcontent.com/pod-product-compliance
Lightning Source LLC
LaVergne TN
LVHW022325200926
848481LV00004B/1432

* 9 7 9 8 2 9 5 5 1 7 6 0 0 *